HOW TO TRANSFER BOOKS
TO KINDLE, KINDLE FIRE
AND KINDLE APP

A Complete Beginners To Pro Guide On
How To Transfer Books To Kindle, Kindle
Fire And Kindle App In 5 Minutes.

I0492821

BY

CHARLES S. MILLS
Copyright©2018

COPYRIGHT

Charles S. Mills

TABLE OF CONTENT

CHAPTER 1

INTRODUCTION

The Kindle is one of the world most famous android tablet with over 3 million users who owns and uses it, with the amazing features it has in it.

Apart from the wild spread popular use of kindle device as an E-reader, it can also perform other functions as transferring books to your Kindle, Kindle Fire and Kindle app. The user interface is a friendly one and it has amazing stuffs in it.

This guide will show you how you can transfer book to Kindle, Kindle Fire and Kindle app; all you need to do is just follow this guide step by step as they are instructed. There are unlimited apps from other sources that can be transferred to your kindle tablet, but if you find this very difficult, there is no needs to worry because I got you cover with the guide lines in this book.

All over the world lot of people haven't been able to use and unleash wonders with its features but this book gives the breakdown of all solution to any problem you might come across.

Thankfully each steps are very easy and simple to follow and understand, that even a novice can master it in a few minutes.

CHAPTER 2

HOW TO TRANSFER BOOKS TO KINDLE, KINDLE FIRE AND KINDLE APP

With Kindle app installed, Kindle users or lovers can read Kindle books on Kindle eReaders, Kindle Fire tablets as well as other smart devices. Generally, if a device that is registered to an Amazon account, all Kindle books that are under that account will be synced automatically when connected to the internet.

However, if we are in need to read non-Amazon books on Kindle or already purchased Kindle books do not appear on the device due to wrong connection, such books will have to be transfer from computer to Kindle manually. In this guide you will be show how books can be transfer to Kindle, Kindle for Android/iPad app from PC or Mac, Kindle paperwhite, Kindle Fire (HD).

1. ENSURE TO PUT YOUR PERSONAL BOOKS IN MOBI FORMAT.

Amazon Kindle are not in support of EPUP books, but they support MOBI (DRM-free) and PDF formats. Personally MOBI is more recommended to PDF because of its flexibility and better for reading

Downloaded books from torrent sites are often EPUP or PDF. If you wish to transfer these books to Kindle for reading, you will have to convert them to Kindle supported MOBI or PDF. In order to do this, you should follow this guide to convert EPUP books to MOBI with free caliber. Ensure to set the output format to MOBI.

If you wish to transfer eBooks purchased from Barnes and Noble, Sony, Kobo or other stores to Kindle, things will be a little hard. It is impossible to convert the formats of books that are DRM protected. Instead, we have to strip the DRM restrictions first. DRM-protected PDF books should also be decrypted because Kindle only supports DRM-free PDF.

2. TRANSFERRING BOOKS TO KINDLE, KINDLE APPS AND KINDLE FIRE(HD) FROM COMPUTER.

Immediately the books are open in MOBI or PDF, they can easily be transfer to PC/Mac to Kindle, Kindle Fire HD, Kindle for Android / iPad app with USB or wirelessly, Paperwhite.

To transfer books to Kindle

a. Get your device connected to computer via USB.

b. Click the drive twice to open it and view the folders.

c. Choose the item you want to transfer (MOBI or PDF).

d. Under Kindle drive copy books to the folder (documents)

e. However, you can as well use the Send-to-Kindle email address.

Below are details information on how to transfer MOBI eBooks to Kindle with USB, and it also works on PDF docs. **Wanting to transfer books to Kindle without a USB cable,** first you should **find out the device email address (mention in each case),** after that you should send your books to the given device/app address.

a. Get MOBI/PDF books transfer to Kindle/Paperwhite eReaders

b. Get MOBI/PDF books transfer to Kindle Fire (HD) tablets

c. Get MOBI/PDF books transfer to Kindle for Android app

d. Get MOBI/PDF books transfer to Kindle for iPad app

e. Get MOBI/PDF books transfer to Kindle without USB

CHAPTER 3

GET MOBI/PDF BOOKS TRANSFER TO KINDLE, KINDLE PAPERWHITE AND OTHER KINDLE E-INK READERS.

In this case I take the Kindle paperwhite for example.

i. With the USB cable get the Kindle device connected to computer. It will be recognized as a Kindle drive.

ii. Under the Kindle drive drag and drop the MOBI books to the "documents" folder.

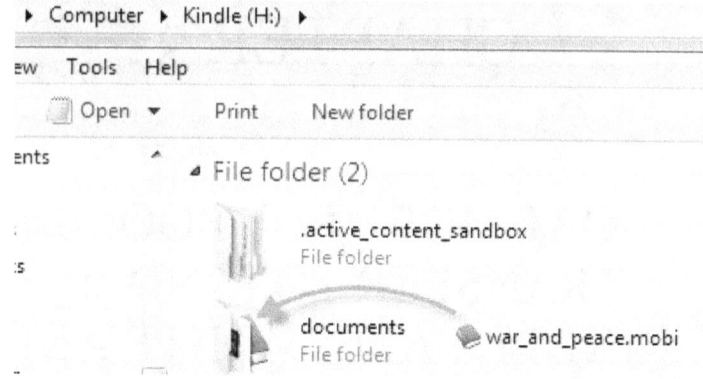

iii. Get the Kindle eject from the computer. The loaded books will display under "Device" shelf.

Send-to-Kindle Email Address of Kindle Paperwhite:

Here you should tap on the menu icon (top right corner) from the home screen, then select "Settings"
"Device options"
"Personalize your Kindle"
"Send-to-Kindle Email".

Send-to-Kindle E-mail

Send documents to your Kindle using the e-mail below.

Go to www.amazon.com/myk to change your settings.

@kindle.com

For Conventional Kindle: click on "Settings" from the home screen menu then use the next and previous page buttons to get the Send-to-Kindle Email. However, this tool enable you transfer books from Kindle / Kobo to Kindle e-ink, even when your books are DRMed or not. Hence, this tool is recommended to people who collected a lot of ebooks files and also had Kindle or Kobo device.

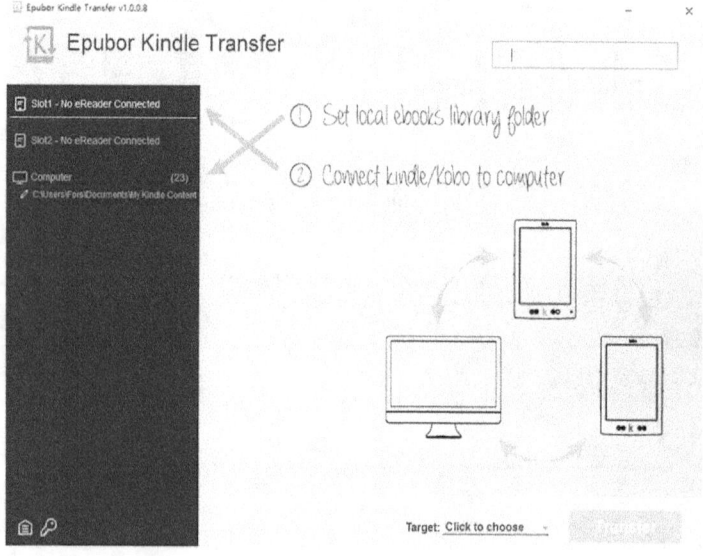

Wirelessly transfer books to Kindle / Paperwhite

CHAPTER 4

GET MOBI/PDF BOOKS TRANSFER TO KINDLE FIRE AND KINDLE FIRE HD

Remember that a USB cable does not come with Kindle Fire HD, failure to have a USB cable, ensure to get the Send-to-Kindle email address of the Kindle Fire and wirelessly transfer MOBI or PDF books to it.
In this case I take the Kindle Fire for example.
Steps On How To Transfer MOBI Or PDF Books To Kindle Fire Or Kindle Fire HD With USB

    i. With the USB cable get your Kindle Fire HD connected to your

computer and it will be
recognized as a Kindle drive.

ii. Under the Kindle Fire HD drive,
copy and paste your MOBI nooks
to the "Books" folder.

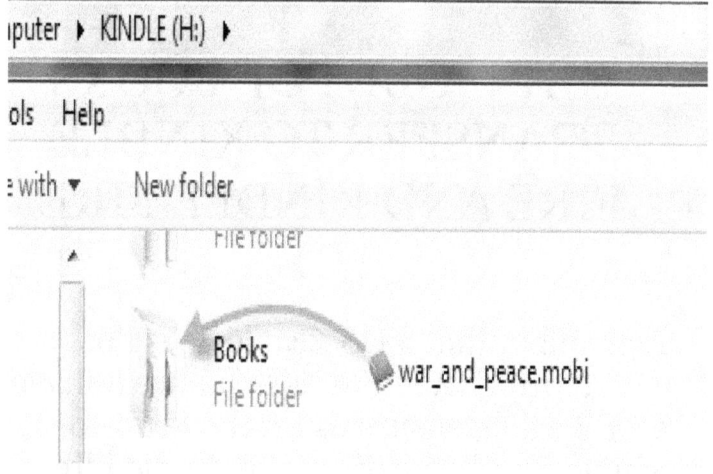

iii. Click on "Books" to check on the transferred books (on the top menu bar) then "Device"

Send-to-Kindle Email Address of Kindle Fire:

Click the top right corner, then choose "More"

"My account" to achieve the device email address.

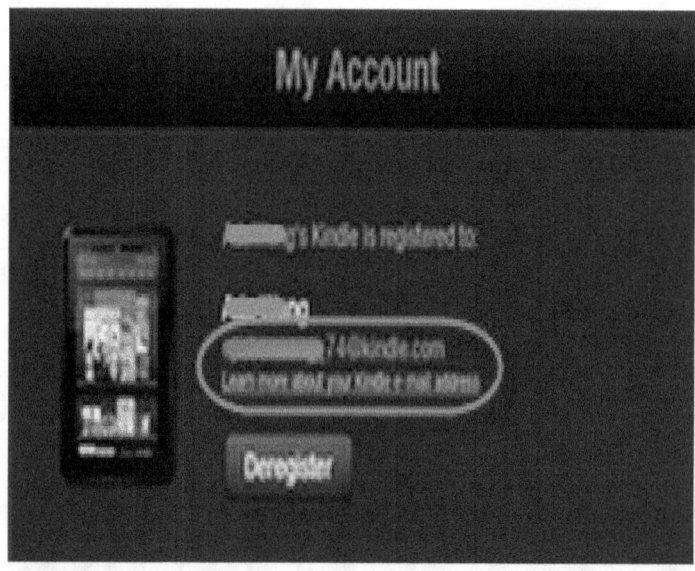

Wirelessly transfer books to Kindle Fire (HD)

CHAPTER 5

GET MOBI/PDF BOOKS TRANSFER TO KINDLE FOR ANDROID APP

In this case I take the Nexus 7 for example

i.   Get your Android tablet or smart phone connected to PC.

ii. Move to the "Kindle" folder of your Android device storage, then copy and paste the MOBI books to that folder.

Computer ▸ Nexus 7 ▸ Internal storage ▸

Tools   Help

File folder

kindle   war and peace.mobi

iii. On the top right corner of the Kindle app click the menu icon, then click "On Device" to check the transferred books.

Send-to-Kindle Email Address of Kindle for Android app:

On the top left corner click the menu icon, then select "Settings" to get the Send-to-Kindle email address.

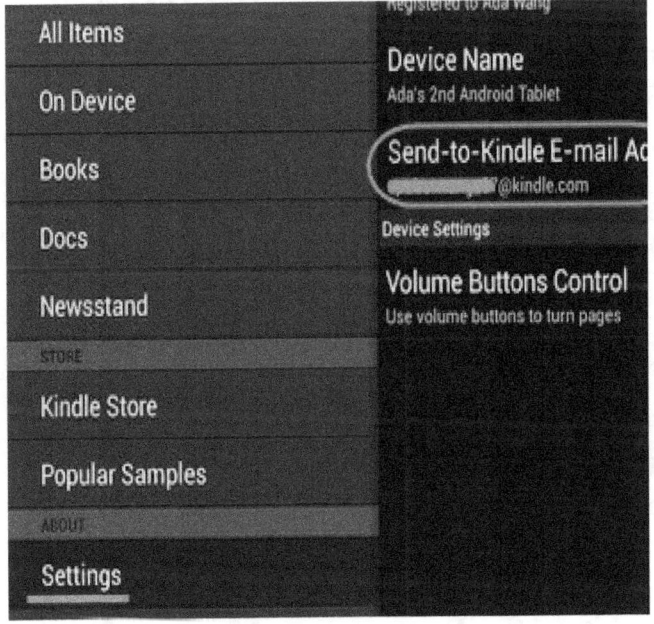

Wirelessly transfer books to Kindle for Android app

CHAPTER 6

GET MOBI/PDF BOOKS TRANSFER TO iPhone / iPad / iPod KINDLE APP

Generally, to directly transfer MOBI books to Kindle iPhone / iPad / iPod app with iTunes has no way yet. However, Kindle email system can be use to send personal MOBI books to the email address of Kindle for iPad app. In this case, with iTunes PDF files can be transferred to iPad (open with iBooks by default) or with Kindle for iPad email address (open with Kindle for iPad app by default).

Send-to-Kindle email Address of Kindle for iPad app:

At the bottom right corner click the gear icon and then click or select "Send-to-Kindle email Address".

Wirelessly transfer books to Kindle for iPhone / iPod / iPad

CHAPTER 7

WIRELESSLY GET MOBI/PDF BOOKS TRANSFER TO KINDLE (WITHOUT USING A USB CABLE)

since the details of the email addresses of the Kindle device or app are known, it will be possible to send DRM-free MOBI or PDF books to Kindle, Kindle Fire (HD), Kindle Paperwhite, Kindle for Android / iPad app wirelessly.

    i. Applying personal (sender) email address to Kindle approved personal document email list.

To get this properly done ensure to follow each step below carefully.

→ Move to "Manage Your Kindle", sign in.

→ Then go to "Your Kindle Account" from the left side menu and

→ Click on "Personal Document Settings". here you should pay much attention to the "Approved Personal Document Email List".

→ Click on "Add a new approved email address".

→ Type in your email address then select "Add Address".

Approved Personal Document E-mail List

To prevent spam, your Kindle will only receive files from the follo

E-mail address

Add a new approved e-mail address

Add a new approved e-mail address

Enter an approved e-mail address.
Tip: Enter a partial address, such as @yourcompany.com, senders.

E-mail address:  @gmail.com

Add Address

    ii. Write an empty email (empty body and empty subject), then you should apply your MOBI or PDF docs and type in the Send-to-Kindle email address to send.

Click on the sync icon on the Kindle device or Kindle app. After few minutes, under "All Items" and "Docs" section (cloud), you will find the sent MOBI and PDF docs there. (however, this screenshot was taken on Nexus 7).

CHAPTER 8

VITAL INFORMATION:

Kindle users should be informed that when you transfer PDF books to Kindle from computer, you will see that the books are not shown under "Books" (on device) section or shelf. This is because PDF books are typically display under "Docs" (on device) shelf.

However, if a Kindle user purchases a book and the purchased books are not auto synced and is due to internet problem, the books can be downloaded to your computer. The downloaded books from Amazon site to your computer are typically AZW3. To transfer the downloaded books to your app or Kindle device follow each step below.

→ Go to Amazon site, click on "Manage Your Kindle".

→ Under "Personal Account"

→ Go to "Actions..."

→ Go to "Deliver to my..."

→ "Download and transfer via USB". To get the books downloaded to your computer or Kindle device.

THE END